Air pollution

Dr. Hemant Pathak

DEDICATION

Dedicated to Shri Sainath Maharaj the all omnipotent of world the most merciful.

CONTENTS

Foreword

Protecting our environment is an essential objective for the any nation, and is also crucial to sustainable development. Presented book; Air pollution; provides a unique insight into the problems our planet faces in terms of clean environment, and what to do about it. This is the only books Written for academics, researchers and practitioners working in environmental pollution and management field, expressed comprehensive and interdisciplinary focus on the ecological issues associated with Air pollution to provide a complete picture of current environmental problem from cause to effect to solution

This book made of 10 years consistently research on environmental issues, makes it ideal source for students, teachers, industrialist, environmental experts and economists.

This book provides an essential guide to researchers, it offers: various causes of pollution; on the challenges and experiences in present scenario.

Simply explained, Air pollution is an important book bringing together diverse viewpoints from academia and environmental agencies and regulators, for all who wish to make a difference in how to plan and manage our Environmental resources.

Dr. Hemant Pathak

M.Sc. (Gold medalist), Ph. D.

Assistant Professor of Engineering Chemistry

Indira Gandhi Govt. Engineering College,

Sagar, MP, India

Glossary

Abatement The reduction or elimination of pollution.

Acid rain The precipitation of dilute solutions of strong mineral acids, formed by the mixing in the atmosphere of various industrial pollutants

Act A law

Acute Exposure One or a series of short-term exposures generally lasting less than 24 hours.

Aerosol Particles of solid or liquid matter than can remain suspended in air from a few minutes to many months depending on the particle size and weight.

Air pollution Toxic or radioactive gases or particulate matter introduced into the atmosphere, usually as a result of human activity.

Air Toxic Any air pollutant for which a ambient air quality standard does not exist that may reasonably be anticipated to cause cancer, developmental effects, reproductive dysfunctions, neurological disorders, heritable gene mutations or other serious or irreversible chronic or acute health effects in humans.

Ash Incombustible residue left over after incineration or other thermal processes.

Atmosphere The 500 km thick layer of air surrounding the earth which supports the existence of all flora and fauna.

Biodiversity A large number and wide range of species of animals, plants, fungi, and microorganisms. Ecologically, wide biodiversity is conducive to the development of all species.

Climate change A regional change in temperature and weather patterns. Current science indicates a discernible link between climate change over the last century and human activity, specifically the burning of fossil fuels.

Combustion Burning. Many important pollutants, such as sulfur dioxide, nitrogen oxides, and particulates (PM-10) are combustion products, often products of the burning of fuels such as coal, oil, gas, and wood.

Contamination The act of polluting or making impure; any indication of chemical, sediment, or biological impurities.

Dust   Solid particulate matter that can become airborne.

Ecosystem   An interactive system that includes the organisms of a natural community association together with their abiotic physical, chemical, and geochemical environment.

Effluent   Municipal sewage or industrial liquid waste (untreated, partially treated, or completely treated) that flows out of a treatment plant, septic system, pipe, etc.

Emission   Release of pollutants into the air from a source. We say sources emit pollutants. Continuous emission monitoring systems (CEMS) are machines, which some large sources are required to install, to make continuous measurements of pollutant release.

Exposure   The concentration of the pollutant in the air multiplied by the population exposed to that concentration over a specified time period.

Fossil fuels   Fuels such as coal, oil, and natural gas; so-called because they are the remains of ancient plant and animal life.

Global warming increase in the average temperature of the earth's surface.

Greenhouse gases Atmospheric gases such as carbon dioxide, methane, chlorofluorocarbons, nitrous oxide, ozone, and water vapor that slow the passage of re-radiated heat through the Earth's atmosphere.

Hydrocarbons   Compounds containing various combinations of hydrogen and carbon atoms. They may be emitted into the air by natural sources (e.g., trees) and as a result of fossil and vegetative fuel combustion, fuel volatilization, and solvent use. Hydrocarbons are a major contributor to smog.

Industrialized Nations whose economies are based on industrial production and the conversion of raw materials into products and services, mainly with the use

countries   of machinery and artificial energy (fossil fuels and nuclear fission).

Kyoto Protocol   An international agreement adopted in December 1997 in Kyoto, Japan. The Protocol sets binding emission targets for developed countries that would reduce the emissions on average 5.2 percent below 1990 levels.

Micro- (µ)   The metric prefix for one millionth of the unit that follows.

Microgram (µg)   One millionth of a gram: $1 \mu g = 10^{-6} g = 0.001$ mg.

Mitigation   Actions taken to avoid, reduce, or compensate for the effects of environmental damage. Among the broad spectrum of possible actions are those that restore, enhance, create, or replace damaged ecosystems.

Monitoring   Periodic or continuous surveillance or testing to determine the level of compliance with statutory requirements and/or pollutant levels in various media or in humans, plants, and animals.

Non-point pollution   diffuse pollution mainly from agriculture or dumping grounds. It is difficult to collect for treatment.

Opacity   The amount of light obscured by particle pollution in the atmosphere. Opacity is used as an indicator of changes in performance of particulate control systems.

Ozone   A gas which is a variety of oxygen. High concentrations of ozone gas are found in a layer of the atmosphere - the stratosphere - high above the Earth. Stratospheric ozone shields the Earth against harmful rays from the sun, particularly ultraviolet B.

Ozone depletion   the reduction of the protective layer of ozone in the upper atmosphere by chemical pollution.

PM-10   particulate matter less than 10 microns in diameter.

Particulate   pollution made up of small liquid or solid particles suspended in the

pollution   atmosphere or water supply.

ppb/ ppm   Units commonly used to express contamination ratios, as in establishing the maximum permissible amount of contaminant in water, land, or air.

Plume   A visible or measurable discharge of a contaminant from a given point of origin that can be measured according to the Ringelmann scale.

Pollutants (pollution)   Unwanted chemicals or other materials found in the air. Pollutants can harm health, the environment and property. Many air pollutants occur as gases or vapors, but some are very tiny solid particles: dust, smoke, or soot.

Point pollution   polluted water from a defined point. It can be collected as industrial or municipal wastewater and treated by what is often called end-of-pipe technology (environmental technology).

Pollution control   The addition of processes, practices, materials, products or energy to waste streams to reduce the risk posed by pollutants and waste before their release to the environment.

Pollution prevention   The use of processes, practices, materials, products, substances or energy that avoid or minimize the creation of pollutants and waste, and reduce the overall risk to human health or the environment

Public health   the health or physical well-being of a whole community.

Reuse   The reemployment of products or materials, in their original form or in new applications, with refurbishing to original or new specifications as required.

Risk assessment   Methods used to quantify risks to human health and the environment.

Smog   A mixture of pollutants, principally ground-level ozone, produced by chemical reactions in the air involving smog-forming chemicals.

Solid waste   non-liquid, non gaseous category of waste from non-toxic household and commercial sources.

Threatened species	species of flora or fauna likely to become endangered within the foreseeable future.
Toxic emissions	poisonous chemicals discharged to air, water, or land.
Toxic waste	garbage or waste that can injure, poison, or harm living things, and is sometimes life-threatening.
Visibility	A measurement of the ability to see and identify objects at different distances. Visibility reduction from air pollution is often due to the presence of sulfur and nitrogen oxides, as well as particulate matter.
Waste	Garbage, trash.

Abbreviations

CO Carbon Monoxide

GHG Greenhouse Gases

NO_2 Nitrogen Dioxide

NO_x Nitrogen Oxides

O_3 Ozone

PAHs Polycyclic Aromatic Hydrocarbons

PM Particulate Matter

PM2.5 Particulate Matter < 2.5 μm in diameter

PM10 Particulate Matter < 10 μm in diameter

ppb parts (of contaminant) per billion (parts of air) by volume

ppm parts (of contaminant) per million (parts of air) by volume

SO_2 Sulphur Dioxide

TSP Total Suspended Particulate

μg/m^3 micrograms (of contaminant) per cubic metre (of air) by weight

VOC Volatile Organic Compound

1. Introduction

Atmosphere is a complex dynamic natural gaseous system that is essential to support life on planet Earth. A substance in the air that can be adverse to humans and the environment is known as an air pollutant.

Air pollution is the introduction into the atmosphere of chemicals, particulates, or biological materials that cause discomfort, disease, death to humans, damage other living organisms such as food crops, natural environment etc..

Exposures to air pollutants have been linked to a wide range of adverse health outcomes, including respiratory and cardiovascular diseases, asthma exacerbation, reduced lung function and premature death.

Japan was the first country in Asia to make a policy decision to use liquefied natural gas for electricity generation in areas where air pollution was already high.The successful implementation of Montreal Protocol shows that it is possible to overcome a serious environmental threat with international cooperation particularly through sharing of responsibilities and resources.

2. Air pollutants

These substances, which include nitrogen oxides, sulfur oxides, carbon monoxide, Total Suspended Particulates (TSP), CO2 and hydrocarbons (such as methane) and volatile organic compounds, are associated with environmental effects such as smog, acid rain and regional haze, and health effects such as respiratory illness.

Nitrogen oxides are removed from the air by rain and fertilize land which can change the species composition of ecosystems. SO_2 is formed when sulphur-containing fuels like coal and oil are burned. In the same vein, NO_2 and greenhouse gases emissions are attributed to fuel combustion for energy generation in motor vehicles, power stations and furnaces.

Industrial air pollution is primarily derived from energy use. Industry consumes over 40 per cent of commercial energy in India and rest south Asia. These pollutants are emitted from a variety of sources, including residential fuel combustion, motor vehicles and agricultural activities. Industrial sources are also major contributors among them, electric utilities, primary metal smelters and cement kilns. Smog and haze can reduce the amount of sunlight received by plants to carry out photosynthesis and leads to the production of tropospheric ozone which damages plants.

Energy efficiency, therefore, is of utmost importance and it needs least cost investments that industrial firms can easily make to reduce air pollution.

Although emissions of air contaminants are trending downward, reductions from sources such as motor vehicles have been partially offset by increases from certain oil and gas industry subsectors attributed to expanded production. Sulfur dioxide and nitrogen oxides can cause acid rain which lowers the pH value of soil.

Energy efficient technologies are implicit in most investment in 'clean technologies', which reduce pollution through reduced inputs and lower pollution intensities.

3. Air pollution in Indian scenario

During the process of industrialization in India air quality declined significantly. where in many urban areas, air pollution greatly exceeds levels considered safe by the World Health Organization (WHO). India is the second largest producer and fourth largest consumer of CFCs in the world. Its production of 23.7 million tons (MT) during 1997 accounted for 16.4 per cent of the world total. During the same year, it consumed 6.7 MT, amounting to about 5.3 per cent of the consumption worldwide.

The worst short term civilian pollution crisis in India was the 1984 Bhopal Disaster. Leaked industrial vapours from the Union Carbide factory, belonging to Union Carbide, Inc., U.S.A., killed more than 24,000 people outright and injured anywhere from 150,000 to 600,000.

India has a commitment under the United Nations Framework Convention on Climate Change (UNFCCC) to monitor its complete greenhouse gases emissions. The greenhouse gases emissions from the industry sector amounted to 405.86 million tons of CO_2, 0.15 million tons of CH_4 and 0.21 million tons of NO_2, which amounted to 412.55 million tons of CO2 equivalent.

The Multilateral Fund for the Implementation of the Montreal Protocol has approved a World Bank project which will assist India in completely phasing out of CFC production by 2010, with production ceilings set for each of the earlier years.

Current particulate matter data in some metro cities in India

Millions of Indians live in metropolitan areas where urban smog, particle pollution, and toxic pollutants pose serious health concerns.

Critical [PM10 > 90 µg/m3]

Guwahati, Patna, Raipur, Delhi, Faridabad, Dhanbad, Nagpur, Bhopal, Indore, Jalandhar, Ludhiana, Jaipur, Howrah, Kolkata High [PM10 61 - 90 µg/m3]

Hyderabad, Chandigarh, Ahmedabad, Panjim, Shimla, Bangalore, Mumbai, Pune, Bhubanshwar Moderate [PM10 31 - 60 µg/m3] Kochi, Shillong, Chennai

Low [PM10 up to 30 µg/m3] Aizwal

Source- CPCB, 2009

4. Classification of Air Pollutants

Air Pollutants can be in the form of solid particles, liquid droplets, or gases. they may be natural or man-made. It classified as primary or secondary.

Primary pollutants are directly produced from a process, such as ash from a volcanic eruption, the carbon monoxide gas from a motor vehicle exhaust or sulfur dioxide released from factories.

Secondary pollutants are not emitted directly. Rather, they form in the air when primary pollutants react or interact. example - ground level ozone make up photochemical smog.

4.1 Primary pollutants

- **Sulfur oxides (SO$_x$)** - sulfur dioxide SO$_2$ is produced by volcanoes and in various industrial processes. coal and petroleum often contain sulfur compounds, their combustion generates sulfur dioxide. oxidation of SO$_2$, in the presence of a catalyst forms H$_2$SO$_4$, and resulted formation of acid rain.

- **Nitrogen oxides (N$_{Ox}$)** - Nitrogen dioxide are expelled from high temperature combustion, produced naturally during thunderstorms by electric discharge. It is one of the most prominent air pollutants.

- **Carbon monoxide (CO)**- It is a colourless, odourless, non-irritating but very poisonous gas. It is a produced by incomplete combustion of fuel such as natural gas, coal or wood. Vehicular exhaust is a major source of carbon monoxide.

- **Volatile organic compounds -** VOCs are greenhouse gases via their role in creating ozone and in prolonging the life of methane in the atmosphere, although the effect varies depending on local air quality. Methane is an extremely efficient greenhouse gas which contributes to enhanced global warming. aromatic compounds viz. benzene, toluene and xylene are suspected carcinogens and may lead to leukemia through prolonged exposure.

- **Toxic metals** like lead, mercury and their compounds.

- **Particulate matter (PM)**, are tiny particles of solid or liquid suspended in a gas. Sources of particulates can be man made or natural.

    Some particulates occur naturally, originating from volcanoes, dust storms, forest and grassland fires, living vegetation, and sea spray.

Human activities, such as the burning of fossil fuels in vehicles, power plants and various industrial processes also generate significant amounts of particulates.

Increased levels of fine particles in the air are linked to health hazards such as heart disease, altered lung function and lung cancer.

- **Chlorofluorocarbons (CFCs)** - harmful to the ozone layer emitted from products currently banned from use.

- **Ammonia (NH$_3$)** - emitted from agricultural processes.

- **Radioactive pollutants** - produced by nuclear explosions, nuclear events, war explosives, natural processes such as the radioactive decay.

4.2 Secondary pollutants

- **Ozone (O$_3$)** is a pollutant, and a constituent of smog formed by chemical reactions in sunlight from air pollutants emitted from fossil fuel burning and industry. At abnormally high concentrations brought about by human activities. Rising concentrations have negative impacts on human health, crop production, tree and vegetation growth.

    The US EPA estimates allowing a ground-level ozone concentration of 65 parts per billion, would avert 1,700 to 5,100 premature deaths nationwide in 2020 compared with the current 75-ppb standard.

ozone can exacerbate acute health problems trigger an asthma attack in someone with asthma.

- **Peroxyacetyl nitrate (PAN)**
- **Persistent organic pollutants (POPs)** are organic compounds that are resistant to environmental degradation through chemical, biological, and photolytic processes. They have potential significant impacts on human health and the environment and capable of long-range transport, bioaccumulate in human and animal tissue, biomagnify in food chains.
- **Photochemical smog** is a kind of air pollution; the word "smog" is a combination of smoke and fog. smog results from vehicular and industrial emissions acted on in the atmosphere by ultraviolet light from the sun to form secondary pollutants that also combine with the primary emissions to form photochemical smog.

The United Kingdom suffered its worst air pollution accident dated on December 4, 1952 Photochemical Smog formed over London. In six days more than 4,000 died, and 8,000 more died within the few months.

5. **Environmental Effects of air pollution**

Air pollution can cause a variety of environmental effects: air pollution can cause a variety of environmental effects:

5.1 **Acid rain**

It is precipitation containing harmful amounts of nitric and sulfuric acids. These acids are formed primarily by nitrogen oxides and sulfur oxides released into the atmosphere when fossil fuels are burned. These acids fall to the Earth either as wet precipitation (rain, snow, or fog) or dry precipitation (gas and particulates).

acid rain damages trees and causes soils and water bodies to acidify, making the water unsuitable for some fish and other wildlife. It also speeds the decay of buildings, statues, and sculptures that are part of our national heritage.

5.2 **Greenhouse Effect**

Emissions of greenhouse gases from human activity are the principal reason for the warming of the earth in recent decades. These gases includes carbon dioxide (CO_2), methane and nitrous oxide, are linked to global climate change.

Industrial energy use is a major source of CO_2 emissions in India. Carbon dioxide emissions cause ocean acidification, the ongoing decrease in the pH of the Earth's oceans as CO_2 becomes dissolved. The emission of greenhouse gases leads to global warming which affects ecosystems in many ways.

5.3 Eutrophication

Eutrophication is a natural process in the aging of lakes and some estuaries. Human activities can accelerate eutrophication by increasing the rate at which nutrients enter aquatic ecosystems. It is a condition in a water body where high concentrations of nutrients (N,P) stimulate blooms of algae, which in turn can cause fish kills and loss of plant and animal diversity.

5.4 Ozone depletion

In the stratosphere, ozone forms a layer that protects life on earth from the sun's harmful ultraviolet (UV) rays. ozone is gradually destroyed by man-made chemicals referred to as ozone-depleting substances, including CFCs, HCFCs, and halons are used in coolants, foaming agents, fire extinguishers, solvents, pesticides, and aerosol propellants.

Thinning of the protective ozone layer can cause increased amounts of UV radiation to reach the Earth, which can lead to more cases of skin cancer, cataracts, and impaired immune systems.

Appearance of Ozone Hole over the Antarctic,1980 was quickly traced back to the rapid increases in the emissions of gases containing chlorine and bromine during preceding decades. UV can also damage sensitive crops, such as soybeans, and reduce crop yields.

5.5 Haze

Haze obscures the clarity, color, texture, and form of what we see. sunlight encounters tiny pollution particles in the air. Some haze-causing pollutants are directly emitted to the atmosphere by sources such as power plants, industrial facilities, trucks and automobiles, and construction activities.

5.6 Global climate change

Greenhouse effect keeps the Earth's temperature stable. humans have disturbed natural balance by producing large amounts of some of these greenhouse gases, including carbon dioxide and methane.

As a result, the Earth's atmosphere appears to be trapping more of the sun's heat, causing the Earth's average temperature to rise a phenomenon known as global warming.

5.7 Crop and forest damage

Crop and forest damage can also result from acid rain and from increased UV radiation caused by ozone depletion. Ground-level ozone can lead to reductions in agricultural crop and commercial forest yields, reduced growth and survivability of tree seedlings, and increased plant susceptibility to disease, pests and other environmental stresses.

5.8 Climate Fluctuations

An increase in the concentration of the GHGs coincided with an increase in the mean surface temperature of the earth of about 4 degrees Celsius.

5.9 Effects on wildlife

Toxics air are contributing to birth defects, reproductive failure, and disease in animals. Toxic pollutants in the air, or deposited on soils or surface waters, can impact wildlife in a number of ways. Humans and animals can experience health problems after exposed to toxics air over time.

5.10 Sea-Level Fluctuation

The expected rise in sea levels due to climate change are still anticipated to range from 0.3 to 0.5 metres by the year 2100 and could present a big challenge to most countries of the region. Concerns have been expressed by the leaders of many small island nations such as, the Maldives, Tuvalu, Kiribati and Tonga, where most land is only a few meters above sea level.

Changes in the sea temperature are also likely to have serious impacts particularly on coral reefs and migratory species of marine life.

Asian countries such as China, India, Indonesia and Bangladesh have substantial parts of their population living close to river deltas, including many of the Megacities of the region, such as Calcutta and Shanghai.

5.11 Economic effects

Air pollution has economic impacts due to increased mortality and illness, the degradation to crops and property and due to tourists avoiding or shortening visits to cities that are heavily polluted. Estimating a monetary value for air pollution impacts is difficult, as it involves estimating non-market costs and values (e.g., health).

According to an estimate by the World Bank (1992), about two to five per cent of all deaths in urban areas in the developing world are

due to high exposures to particulates.

5.12 Effects on ecosystems

The damage caused by air pollution on ecosystems may be less obvious and more difficult to quantify, Air pollution can cause damage to plants and animals, to aquatic and terrestrial ecosystems, impacting on biodiversity and damaging valued habitats.

Deposition of sulphur and/or nitrogen can cause increased acidity, and when critical loads8 for acidity levels are exceeded, ecosystem damage may occur.

5.13 Health effects

The World Health Organization states that 2.4 million people die each year from causes directly attributable to air pollution, with 1.5 million of these deaths attributable to indoor air pollution.

Air pollution can harm us when it accumulates in the air in high enough concentrations. air pollution can cause cancer and damage to the immune, neurological, reproductive, and respiratory systems. air pollution can cause cancer and damage to the immune, neurological, reproductive, and respiratory systems. People exposed to high enough levels of certain air pollutants may experience:

- ❖ Lung and heart problems, such as asthma

- ❖ Irritation of the eyes, nose, and throat

- ❖ Coughing, chest tightness, wheezing and breathing difficulties

- ❖ Risk of heart attack

Worldwide more deaths per year are linked to air pollution than to automobile accidents. Causes of deaths include aggravated asthma, emphysema, lung and heart diseases, and respiratory allergies. The US EPA estimates that a proposed set of changes in diesel engine technology could result in 12,000 fewer premature mortalities, 15,000 fewer heart attacks and 6,000 fewer emergency room visits by children with asthma.

The US EPA projects the stricter standard would also prevent an additional 26,000 cases of aggravated asthma, and more than a million cases of missed work or school.

Diesel exhaust (DE) is a major contributor to combustion derived particulate matter air pollution. In several human experimental studies, using a well validated exposure chamber setup, DE has been linked to acute vascular dysfunction and increased thrombus formation. This serves as a plausible mechanistic link between the previously described association between particulates air pollution and increased cardiovascular morbidity and mortality.

6. Air pollutant emission factors

This value create to relate the quantity of a pollutant released to the ambient air with an activity associated with the release of that air pollutant. These factors are usually expressed as the weight of pollutant divided by a unit weight, volume, distance, or duration of the activity emitting the pollutant (e.g., kilograms of particulate emitted per tone of coal burned). This factors useful for estimation of emissions from various sources of air pollution.

7. Control of Air pollution and Reduction efforts

Air Pollution control is a big problem associated with developing countries like India. All anti-pollution laws and measures, become ineffective, in the absence of proper monitoring system. This can be accomplished through regular monitoring of big industries by a government agency. It is essential to control of emissions and effluents into air under environmental management.

Without pollution control, the waste products from consumption, agriculture, heating, mining, manufacturing, transportation and other human activities, whether they accumulate or disperse, will degrade the environment. pollution prevention and waste minimization are also necessary for pollution control .

There are various air pollution control technologies available to reduce air pollution. Efforts to reduce pollution from mobile sources includes increased fuel efficiency, conversion to cleaner fuels such as bio ethanol, biodiesel and use of electric vehicles.

8. Air pollution Control devices

The following items are commonly used as air pollution control devices by industry and transportation devices. They can either destroy contaminants or remove them from an exhaust stream before it is emitted into the atmosphere.

8.1 Particulate matter control

I. Mechanical collectors

II. Electrostatic precipitators

III. Baghouses

IV. Particulate scrubbers

8.2 NOx control

I. NOx scrubbers

II. Low NOx burners

III. Selective catalytic reduction (SCR)

IV. Selective non-catalytic reduction

V. Exhaust gas recirculation

VI. Catalytic converter

8.3 Volatile Organic Compound

I. Cryogenic condensers

II. Adsorption systems

III. Biofilters and Flares

IV. Thermal oxidizers

V. Catalytic converters

VI. scrubbing

VII. Vapor recovery systems

8.4 Acid Gas/SO_2 control

I. Wet scrubbers

II. Dry scrubbers

III. Flue-gas desulfurization

8.5 Mercury control

I. Sorbent Injection Technology

II. Electro-Catalytic Oxidation (ECO)

III. K-Fuel

9. Prevention of Air pollution

Prevention of air pollution can be achieved by reduction or elimination of wastes and pollutants at their sources. Natural resources have to be prolonged to their completely use to maintain the aim for continual economic growth and lessen environmental impacts. This involves reducing wastage in operations, utilizing waste products through recycling and recovery practices to further ensure the long-term availability and usefulness of natural resources.

For all the pollution that is avoided in the first place, there is that much less pollution to manage, treat, dispose of, or clean up. This goal may be achieved by some activities such as:

I. Implementing better housekeeping practices to minimize leaks and fugitive releases from manufacturing processes. Redesigning products to cause less waste or pollution during manufacture, use, or disposal altering production processes to minimize the use of toxic chemicals.

II. To reduce energy consumption Pollution prevention within industry generally receives the most attention. Replacing coal and oil with natural gas where feasible can improve air quality. Replacing coal and oil with natural gas where feasible can improve air quality.

III. Suspended particulates matter and sulphur dioxide, it is possible to install devices like electrostatic precipitators and scrubbers that can reduce emissions by 90 per cent or more.

IV. To reduce emissions of carbon dioxide from a power plant through improving the efficiency of the boiler keeping the environment clean and managing the wastes with the guide lines of respective Government. The Exhausts from the Automobiles and workshop machinery should be controlled. Repair and replacement of leaking and malfunctioning equipment.

V. Compact fluorescent lighting, increasingly used in offices, is another example of a technology that can reduce energy use by more than 50 per cent, while providing the same amount of lighting.

VI. The ISO standards must be followed strictly for Industrial usage.

VII. Emissions of CO_2 as well as of traditional air pollutants can be reduced by fuel substitution.

VIII. To used Eco-friendly means like bicycle, bike etc. Must used public transportation means like bus for routine jobs. Administration must promoted car pool to office and back. Methane from landfills, coal, mines, and oil production, is beginning to be utilized as energy in a number of countries in the region like India.

IX. Reduce the use of aerosols in the household. Promote the afforesting. Switch-off all the lights and fans when not required. Promoted to sharing of room with others when the air conditioner, cooler or fan is on.

There is a linear relationship between the amount of energy used and the emission of air pollutants. Utilizing equipment that produces the same output while requiring less energy is frequently one of the most cost-effective approaches to improving air quality. Newer refrigerators and computers, for example, use substantially less energy than their predecessors. One measure of energy efficiency is the ratio of energy use to Gross Domestic Product (GDP).

Environmental management assumes paramount importance in this perspective to address the numerous issues relating to pollution control, safety etc., and to minimize the degradation of the environment on account of developmental activities.

10. Future Strategies

Efforts of the global community to reduce emissions of greenhouse gases and other air pollutants continue. Due to the long life of GHGs in the atmosphere a degree of climate change is very likely to occur. It must be prudent to begin planning to minimize the possible adverse effects in some part of world that may be particularly vulnerable,

11. The Air (control of pollution prevention and) act, 1981 in India

An Act to provide for the prevention, control and abatement of air pollution, for the establishment, with a view to carrying out the aforesaid purposes, of Boards, for conferring on and assigning to such Boards powers and functions relating thereto and for matters connected therewith.

Whereas decisions were taken at the United Nations Conference on the Hum an Environment held in Stockholm in June, 1972, in which India participated, to take appropriate steps for the preservation of the natural resources of the earth which, among other things, include the preservation of the quality of air and control of air pollution;

and whereas it is considered necessary to implement the decisions aforesaid in so far as they relate to the preservation of the quality of air and control of air pollution;

be it enacted by Parliament in the Thirty-second Year of the Republic of India

12. References

1. http://www.toronto.ca/health/hphe

2. WHO (2009). Global health risks: Mortality and burden of diseases attributable to selected major risks. Geneva, World Health Organization
http://www.who.int/healthinfo/global_burden_disease/GlobalHealthRisks_report_full.pdf.

3. WHO (2008). Air quality and health. Geneva, World Health Organization
http://www.who.int/mediacentre/factsheets/fs313/en/index.html.

4. WHO (2006). Air quality guidelines—global update 2005. Particulate matter, ozone, nitrogen dioxide and sulfur dioxide. Copenhagen, World Health Organization Regional Office for Europe
http://www.euro.who.int/__data/assets/pdf_file/0005/78638/E90038.pdf.

5. WHO (2006). WHO global air quality guidelines for particulate matter, ozone, nitrogen dioxide and sulfur dioxide—Global update 2005: Summary of risk assessment. Geneva, World Health Organization
http://whqlibdoc.who.int/hq/2006/WHO_SDE_PHE_OEH_06.02_eng.pdf.

6. U.S. Environmental Protection Agency, Office of Pollution Prevention. Pollution Prevention 1991: Progress on Reducing Industrial Pollutants. EPA 21P-3003. Washington: U.S. EPA, October 1991, pp. 6–7.

7. Harry Freeman et al. "Industrial Pollution Prevention: A Critical Review." Journal of Air and Waste Management 42, no. 5 (May 1992): 619–620.

8. U.S. EPA, Pollution Prevention 1991, pp. 6–7.

9. Central Pollution Control Board (2012). National Summary, available at http://www.cpcb.nic.in/FinalNationalSummary.pdf.

10. World Bank (2010). World Development Report: Development and Climate Change, available at http://wdronline.worldbank.org//worldbank/a/c.html/world_development_report_2010/abstract/WB.978-0-8213-7987-5.abstract.

11. Woodruff, T. J., Parker, J. D. & Schoendorf, K. C. (2006). Fine Particulate Matter (PM2.5) Air Pollution and Selected Causes of Post neonatal Infant Mortality in California, Environmental Health Perspectives. 114(5), pp. 786–790.

12. World Bank, (2002). What Do We Know About Air Pollution?—India Case Study, Urban

Air Pollution, South Asia Urban Air Quality Management Briefing Note No. 4, pp. 1-4.

13. World Health Organization (WHO), (2010a). Air Quality: Volcanic Ash Cloud over Europe. Retrieved from http://www.euro.who.int/en/what-we-do/health-topics/environmental-health/air- quality/volcanic-ash-cloud-over-europeWoodruff,

14. http://en.wikipedia.org/wiki/Air_pollution

ABOUT THE AUTHOR

Dr. Hemant Pathak held positions as Assistant Professor in the department of chemistry, Govt. Indira Gandhi Engineering College, Sagar, MP, India. He had extensive experience in teaching, research and administrative management.

Dr. Pathak received his Ph.D. degree in chemistry from Dr. Hari Singh Gour Central University, Sagar, India and M.Sc. Gold medalist from Jiwaji University, Gwalior. He has published 13 books and more than 50 research papers in reputed International and National journals and received several awards. He is a member of editorial boards and reviewer boards of several international journals and societies. His area of specialization includes Engineering Chemistry and Environmental Pollution management.